SOLDERING IRON

HOW TO SOLDER BEGINNERS GUIDE

LEO VIC B.N.

CONTENTS

CHAPTER ONE 3

Introduction for iron soldering 3

CHAPTER TWO 5

Types of Soldering Irons............................... 5

CHAPTER THREE 8

Choosing a Soldering Iron......................... 8

CHAPTE FOUR 10

Soldering Iron Maintenance........................ 10

CHAPTER FIVE 13

Safety Tips for soldering iron 13

CHAPTER SX 15

BEGINNERS PROJECTS 15

CHAPTER ONE

Introduction for Iron Soldering

Iron soldering is a system of becoming a member of two or extra portions of metallic collectively with the aid of melting and flowing a filler metallic into the joint. The filler steel has a decrease melting factor than the work piece metal, permitting it to soften and float into the joint to shape a robust bond. Iron soldering is a necessary talent in many areas of manufacturing, electronics and plumbing, and is used to create permanent seals and connections that would in any other case be hard or not possible to achieve. Iron soldering requires the use of a soldering iron and different specialised tools, such as flux and soldering paste. It is important to make sure that the soldering iron used is gorgeous for the job, and that the acceptable security precautions are taken. Iron soldering is a distinctly easy process, however it requires exercise and ability to attain first-class results. It is vital to understand all of the factors concerned in the technique and to exercise on scrap steel earlier than trying to solder an necessary project. Once mastered, iron soldering can be a very lucrative and pleasing process. With the proper equipment and techniques, you can create strong, dependable connections and repairs that will final for years.

CHAPTER TWO

Types of Soldering Irons

1. Corded Irons: These are the most frequent kind of soldering iron and are normally the least expensive. They are powered by means of mains electricity and can be used for a range of soldering jobs.

2. Gas Soldering Irons: These are powered by using both butane or propane and are perfect for jobs that require portability. They are also beneficial in areas the place there is no get right of entry to an electrical outlet.

3. Battery-Operated Soldering Irons: These are powered via rechargeable batteries and are perfect for use in far flung places or on the go.

4. Stationary Soldering Irons: These are the most effective kind of soldering iron and are normally used for extra elaborate soldering jobs. They are typically the most costly kind of soldering iron.

5. Temperature-Controlled Soldering Irons: These are the most superior kind of soldering iron, permitting the

consumer to exactly control the temperature of the iron. This is best for refined soldering jobs.

6. Soldering Guns: These are typically used for large soldering jobs, such as wiring and circuit boards. They are powered by means of electricity and can attain excessive temperatures quickly.

7. Solder Pots: These are perfect for large soldering jobs, such as soldering a couple of wires together. They are powered with the aid of electricity and can be used to soften massive quantities of solder at once.

8. Solder Fumes Extractors: These are used to assist minimize the quantity of hazardous fumes launched at some point of soldering. They are usually powered through electrical energy and are usually used in areas the place air flow is limited.

9. Hot Air Soldering Stations: These are normally used for greater complex soldering jobs and are powered by means of electricity. They use a aggregate of warm air and solder to melt the elements together.

10. Ultrasonic Soldering Irons: These are the most superior kind of soldering iron and use ultrasonic waves to soften the solder.

They are generally used for very tricky jobs and are commonly the most high-priced kind of soldering iron. No depend what kind of soldering iron you are using, security ought to usually be the pinnacle priority. Always put on protection glasses and a respirator when soldering, as the fumes can be hazardous to your health.

CHAPTER THREE

CHOOSING A SOLDERING IRON

When selecting a soldering iron, there are countless matters to consider. First, the dimension and kind of soldering iron you want relies upon on the initiatives you graph to use it for. If you layout to do small refined work on circuit boards or jewelry, you will desire a smaller iron with a best tip. For large projects, such as auto wiring, you will want a large iron with a wider tip. Secondly, consider the wattage and temperature vary of the iron. A soldering iron with a greater wattage will warmth the tip quicker and be better for large projects, whilst a decrease wattage iron can also be higher for smaller projects. Finally, think about the aspects and accessories that come with the iron. Some irons come with temperature control, replaceable tips, and different points that can make soldering less complicated and greater efficient. Once you have taken these elements into consideration, it is essential to pick a soldering iron that is of desirable best and from a official manufacturer. This will make certain that your iron will closing longer and be greater reliable.

CHAPTE FOUR

SOLDERING IRON MAINTENANCE

1. Clean the soldering iron tip regularly. To smooth the tip, you can use a copper sponge, brass sponge, moist sponge, or a extraordinary soldering iron tip cleaner.

2. Tin the soldering iron tip. To do this, vicinity the solder wire onto the tip and permit it to soften and cowl the tip completely. This will guard the tip from corrosion and oxidation.

3. Replace the soldering iron tip when it will become worn or damaged.

4. Store the soldering iron in a secure location when now not in use to forestall accidents.

5. Unplug the soldering iron when now not in use to keep electricity and keep away from overheating.

6. Inspect the soldering iron wire usually for any symptoms of put on and tear and exchange it if necessary.

7. If the soldering iron is used for prolonged durations of time, enable it to cool down earlier than storing it away.

8. Ensure that the soldering iron is used in accordance to the directions provided.

9. Always use protection measures when working with a soldering iron such as sporting protection goggles and the usage of a soldering iron stand.

10. Always dispose of soldering iron factors in an environmentally pleasant way.

11. Use a soldering iron with the right wattage for the job at hand.

12. Use the right solder for the job at hand.

13. Keep the soldering iron tip easy and free of debris.

14. Use a damp material to wipe away any extra solder from the soldering iron tip.

15. Do now not depart the soldering iron unattended whilst it is plugged in.

16. Always unplug the soldering iron earlier than altering the tip or making any repairs.

17. If the soldering iron is no longer being used for a lengthy duration of time, save it in a dry, cool place.

18. Never depart the soldering iron plugged in for prolonged durations of time.

CHAPTER FIVE

SAFETY TIPS FOR SOLDERING IRON

1. Wear security glasses to defend your eyes from the brilliant mild produced through the soldering iron.

2. Make positive that the soldering iron is right grounded earlier than use.

3. Always use a soldering iron stand and in no way depart it unattended.

4. Never contact the soldering iron tip or the molten solder.

5. Be positive to maintain the soldering iron tip easy and free from oxidation.

6. Make certain that the work vicinity is nicely ventilated.

7. Be certain to use the ideal wattage for the job.

8. Never go away the soldering iron plugged in when now not in use.

9. Disconnect the iron from the strength supply earlier than altering recommendations or cleansing the tip.

10. Keep the soldering iron away from young people and pets.

11. Unplug the soldering iron earlier than cleansing it and permit it to cool earlier than handling.

12. Be positive to use a damp sponge to smooth the tip of the soldering iron after every use.

13. If the soldering iron produces smoke or sparks, unplug it without delay and investigate it for damage.

14. When finished, unplug the soldering iron and save it in a cool, dry place.

15. Dispose of the soldering iron precise when finished.

16. Be certain to study and comply with the manufacturer's guidelines for perfect use and safety.

17. Never strive to restore a broken or malfunctioning soldering iron.

CHAPTER SX

BEGINNERS PROJECTS

Circuit board

Materials:

1. Solder

2. Circuit board

3. Resistors

4. Capacitors

5. Diodes

6. Transistors

7. Wires

8. Heat sink

9. Solder flux

Instructions:

1. Begin through laying out the elements of the circuit board onto the board, making certain to right discover every aspect and its purpose.

2. Connect the aspects with solder flux and solder. When soldering, make certain to use a warmness sink to guard the factors from overheating.

3. Once the factors are connected, use wires to join them in the perfect order to entire the circuit.

4. Once all of the aspects are in place, take a look at the circuit board to make positive it is working properly.

5. Finally, bundle the circuit board with the excellent shielding substances to make certain it is protected in transit.

Damaged rings item

Materials:

-Jewelry glue

-Scissors

-Tweezers

-Fine-tipped paintbrush

-Small bowl

-Paper towels

Instructions:

1. Inspect the damaged rings item, become aware of the place it desires to be reattached, and decide if any portions are missing.

2. If necessary, use the scissors to reduce off any extra portions of the rings item.

3. Apply a small quantity of rings glue to a paper towel.

4. Use the fine-tipped paintbrush to observe the glue to the damaged components of the earrings item.

5. Place the earrings object in a small bowl and enable the glue to dry for about 15 minutes.

6. Once the glue has dried, use tweezers to cautiously reattach the two damaged portions of jewelry.

7. Allow the earrings object to dry definitely earlier than wearing.

8. Store the rings object in a protected vicinity when now not in use.

Creating customized mild installation

Materials:

- Light bulbs (appropriate wattage for your installation)
- Electrical wiring
- Switches
- Extension cords
- Mounting hardware (screws, nails, etc.)
- Tape (electrical, duct, etc.)
- Wire cutters

- Electrical tape

Instructions:

1. Decide on the placement and measurement of your light installation. This will assist you decide the variety and kind of mild bulbs you'll need.

2. Gather the fundamental substances and tools.

3. Measure and mark the spot the place you will mount your mild installation.

4. Install the mounting hardware to the wall or ceiling.

5. Connect the electrical wiring to the mild bulbs.

6. Connect the switches and extension cords to the wiring.

7. Test the mild set up to make positive it is working properly.

8. Secure any unfastened wires with tape.

Customized telephone cellphone case

Materials:

-Your telephone phone

-Plastic case for your phone

-Scissors

-Colored paper

-Glue

-Markers

-Paint pens

-Rhinestones (optional)

Instructions:

1. Take your telephone and the plastic case for it and measure the case to the dimension of your phone.

2. Cut the coloured paper to dimension and glue it to the plastic case.

3. Use markers, paint pens, and rhinestones to beautify your case.

4. Let the glue dry and then insert your telephone into the case.

LED mild strip

Materials:

-LED Light Strip

-Power Supply

-Extension Cable

-Remote Control

-Mounting Tape

-Screws

Instructions:

1. Decide the place you prefer to mount the LED mild strip. Make positive it is on a flat and easy surface.

2. Cut the LED mild strip to the favored length, if needed.

3. Peel off the mounting tape and stick it onto the preferred surface.

4. Connect the energy grant to the LED mild strip.

5. Connect the extension cable to the strength supply, if needed.

6. Connect the faraway manipulate to the electricity supply.

7. Secure the LED mild strip with the supplied screws.

8. Turn the strength provide on and use the far flung manage to regulate the brightness and shade of the LED mild strip.

Crafting a customized amplifier

Materials:

1. Soldering iron

2. Solder

3. Pre-amplifier circuit board

4. Power amplifier circuit board

5. Heat-shrink tubing

6. Aluminum casing

7. Screws and nuts

8. Audio enter and output connectors

9. Potentiometers

Instructions:

1. Start via soldering the pre-amplifier circuit board together. Make certain all connections are secure.

2. Now solder the electricity amplifier circuit board together. Again, make certain all connections are secure.

3. Connect the pre-amplifier and energy amplifier circuit boards collectively with heat-shrink tubing.

4. Place the circuit boards in the aluminum casing and impervious them with screws and nuts.

5. Connect the audio enter and output connectors to the circuit boards.

6. Finally, connect the potentiometers to the circuit boards. This will enable you to regulate the volume, bass, and treble.

Constructing a customized radio

Materials:

1. Soldering Iron

2. Wire Strippers

3. Wire Cutter

4. Multimeter

5. Circuit Board

6. Resistors

7. Capacitors

8. Transistors

9. Diodes

10. Antenna

11. Radio Parts

Instructions:

1. Gather all the fundamental substances and equipment wished to assemble your customized radio.

2. Begin with the aid of planning out the layout of your radio. First, figure out on the points you prefer your radio to have and the factors you will want to construct it.

3. Once you have a sketch in place, start soldering the factors onto the circuit board. Make positive to use the suited soldering techniques and take your time to make sure a excellent connection.

4. Once the factors are in place, join the wires to the board. Use the multimeter to check the connections for accuracy.

5. Once the wiring is complete, take a look at the radio by way of connecting the antenna and plugging it into a energy source. Tune the radio to a local station and regulate the elements as needed.

6. Finally, region the case round the radio and invulnerable the factors in place. You now have a custom, working radio!

Building a DIY robot

Materials:

1. Microcontroller

2. Motors and Wheels

3. Batteries

4. Cables

5. Chassis

6. Wiring

7. Soldering iron

8. Sensors

9. Electrical components

10. Programming language

Instructions:

1. Gather the essential substances and tools.

2. Design the robotic through choosing the splendid chassis, motors and wheels, and sensors.

3. Assemble the robotic by means of connecting the chassis, motors and wheels, batteries, and sensors the usage of cables, wiring, and soldering.

4. Program the robotic the usage of a programming language.

5. Test the robot's performance and make changes as necessary.

www.ingramcontent.com/pod-product-compliance
Lightning Source LLC
Chambersburg PA
CBHW081811240526
45465CB00032BA/2806